Animal Architecture

edited by
Roger Caras

foreword by
Roger Tory Peterson

Westover
Publishing Company
An affiliate of Media General, Richmond, Va.

Prepared in cooperation with
Photo Researchers, Inc.,
New York, New York.

Book design by Sylvan Jacobson

Copyright® 1971 by Westover Publishing Co. First Edition Library of Congress Catalogue Card Number 78-161076 All rights reserved.

This book, or parts thereof, must not be reproduced in any form without written permission from Westover Publishing Co.

SBN 0-87858-011-5

INTRODUCTION

No aspect of the behavior of animals has taught us more than the architectural skills they exhibit. We have watched and we have emulated them.

The cell construction in a beehive is as sophisticated as anything we have yet come up with.

The spider web is as gracefully powerful and as completely wonderful as our most advanced suspension bridge.

And a giant termite nest is more complex than our most complicated building.

I doubt very much that a time will ever come when man will not be amazed by the fact that animals build through instinct. While our young architects struggle through their courses in higher math, the bees and the ants, the beavers and the birds are out there doing it all—without benefit of education.

The skills of animal architects come built into the original genetic package. That is the other world. We with our massive and complex intellects are aware of everything and must learn everything.

The animal builders presumably feel nothing emotionally, and learn practically nothing architecturally. These are two ways in which nature has equipped her children to face the problem of survival. It may take a few more million years to see which is the better system. Of course, if we are not careful we might bring that showdown forward in time.

Roger Caras

FOREWORD

Animals of nearly every category build things. Some are highly intricate, such as the skillfully woven nests of weaver birds, the hexagonal cells of honey bees, and the extraordinary webs of spiders. Others are very simple such as the burrows of many mammals, birds and insects.

Man, of course, is the master architect, not committed to a set method or design. But he must learn his craft; it is not innate or instinctive as it is with his lesser associates. A weaver bird that has been reared in captivity without parents and which has never seen a nest of its own kind will construct, very competently, the right kind of nest when it is furnished with nesting materials.

How it all started is lost in the concealing mists of time, but the variety, the radiation of form, that we see today in animal architecture, is the result of natural selection in the Darwinian sense: it is inherent in the evolutionary process. An individual born with a slight quirk in its makeup may do its thing just a bit differently—an extra fillip here or there. If the departure has survival value, it might be passed from generation to generation in the genes.

The other animals with which we share the earth build things for many of the same reasons that we do. However, most birds build their nests solely as temporary cradles for their young, nothing more. For many mammals a nest or a burrow may act both as a nursery and as a shelter. Spiders and some insects devise hiding places and traps. It all pays off in survival.

Roger Tory Peterson
Old Lyme, Connecticut

White-Footed Mouse

Some animals do not necessarily display their architectural capabilities. Apparently not overwhelmed with the need to build the way many birds are, some mammals will take a short cut and accept a cavity or other natural formation. Still, they display their territorial instincts. It is not the thing that counts even when animals do build—it is the place and what the place can mean. Of course, even the white-footed mouse that takes life easy and accepts a cavity will work to line it with down, fur and feathers—not to mention a few dozens of materials that are sometimes used. After all, even a cavity should be comfortable.

Bird at Nest

Whatever else a bird's nest may be, it is not an isolated phenomenon. It is part of an extremely complex pattern of behavior. It is tied up with body changes in both the male and the female, and it is involved with the seasons. It comes about after territorial instincts have been satisfied (and that involves bird song and some display). It is all part of or immediately follows courtship. Mating, of course, is part of it and so is egg laying. Then comes brooding of the eggs, raising of the chicks. It is amazingly complex. And it works. Nest building or claiming is only a small part of it. But, like every other part, it is essential to the success of the whole system of reproduction.

Elf Owl

"By comparison with other birds, owls are not great homemakers. More often than not they rely upon the endeavours of other species, because they are basically hole or crevice nesters. Old woodpeckers' holes are frequently used . . . the tiny elf owl depends upon Gila woodpeckers and Mearns' gilded flickers to bore out nesting sites in the giant Saguaro cactus. Indeed, owls and woodpeckers may even use the same cavity, the owls sleeping by day and moving out as the Gila woodpeckers come home to roost."

From *Owls: Their Natural and Unnatural History*
by John Sparks and Tony Soper
Taplinger Publishing, N.Y. 1970

Female Wolf Spider

"The female wolf spider is, in the fashion of her sex, a creature of variable temper. Notorious for her rapacious activities, she nevertheless displays a solicitude for her eggs and young that can scarcely be matched by any other spider. The mother *Pardosa* . . . encloses her eggs in a carefully molded spherical bag, attaches the sac to her spinnerets and drags it around with her wherever she goes. It makes no difference that it is often as large as she is; this egg bag is a precious thing to her; she will defend it with her very life, and will fight viciously to retain it."

From *American Spiders*
by Willis J. Gertsch
D. Van Nostrand Co., Princeton, N.J. 1949

House Martin   Weasel Least-White

There are two basic forms of nests. The house martin above has built a nest from mud balls. (Both members of a mated pair actually participated.) The weasel (above) and the ground squirrel (right) have excavated their nests. In the case of the bird, earth was taken from one place, moved to another and a structure created from it. In the case of the two mammals, the task was one of earth removal to create a cavity.
One could say that one is a concave action, the other convex. But, in nature, things do not hold comfortably still just so we'll not have to think. Some birds and mammals reverse these roles. The birds burrow and the mammals build a nest. There is another basic difference in the examples illustrated here. The bird will *not* use the nest for a den—only as a nursery. The mammals will in all probability each use their excavation as nursery, larder, daily retreat and hibernaculum.

Striped Ground Squirrel

Brown Bats

There is a matter of place. To the little brown bats in the attic (left) a place is what matters. There is no creation involved, no building, no adapting. Although mammals, and therefore of higher-than-bird intelligence, these small creatures take a dark retreat and claim it en masse. They take things as they find them. For the red-eyed vireo (below) life is more complicated. The territory was established according to certain basic standards—height above the ground, thickness of cover, proximity to food sources and water, and then the site was altered. Vegetable matter was carried to the selected place and a structure created to accommodate the species' needs. One big difference, of course, is the method of reproduction. The bats will have their young develop within their bodies. The birds will have their young mature in eggs outside of the mothers' bodies. In a sense, then, a bird's nest is not only a location, and a structure, but a kind of womb.

Red-Eyed Vireo at Nest

Chipmunk

"The year before, the chipmunk had worked long and hard to excavate her underground home. She had dug down almost straight for nine or ten inches, then continued the two-inch-in-diameter tunnel on a twisting, sloping course that went nearly three feet below the ground surface. As she dug, loosened soil was kicked behind her in the tunnel. Now and then she paused, turned about and used nose and forepaws to push the soil outside the burrow opening. Chipmunks are always careful not to mark their burrow entrances by mounds of excavated soil."

From *Squirrels of North America*
by Dorcas MacClintock
Van Nostrand Reinhold, N.Y. 1970

Soft Coral
"The soft coral, or sea finger, secretes no limy cups as the distantly related stony, or reef, corals do, but forms colonies in which animals live embedded in a tough matrix strengthened with spicules of lime. Minute though the spicules are, they become geologically important where, in tropical reefs, the soft corals, or Alcyonaria, mingle with the true corals. With the death and dissolution of the soft tissues, the hard spicules become minute building stones, entering into the composition of the reef."
    From *The Edge of the Sea*
    by Rachel Carson
    Houghton Mifflin, Boston 1955

White or Fairy Tern

Even birds that live in the vicinity of water vary greatly in
nesting habits. Here are three good examples. The fairy tern
(left), not to be confused with a black-capped species
found in Australia and also called a fairy tern, does almost
nothing in the way of nest construction. The flamingo nest,
below, represents a fair construction effort using mineral
matter to elevate the egg platform above the level of the lake
or marsh. Also below, a pair of black swans have accumulated,
hundreds of pounds of mud and vegetable debris to build a
monumental nesting platform surrounded by a natural moat.
Each species must find its own special way in this world.
Each species must utilize its surroundings in the way best
suited to species survival. The only common denominator
is survival.

Flamingo Nests   **Black Swan**

Galapagos Tortoise

Badger

Many different kinds of animals dig into the receptive earth in order to complete their life cycles. And they do so in diverse ways. Here are two good examples. The badger, a highly sophisticated mammal, a carnivore and type of giant weasel, digs deep burrows in which it sleeps, hides from the weather and there it bears its young. The burrow is the center of its life. The Galapagos tortoise, on the other hand, does not live underground. It does, however, bury its eggs. It depends on the warm soil to brood its eggs and hatch its young. In both species survival depends on the ability to excavate a cavity. The uses to which these holes in the ground are put vary greatly, however.

Rookery of Adelie Penguins, Antarctica
For the Adelie penguin, nest architecture is as much a matter of place and ritual as it is of building technique. The nest of this little Antarctic penguin consists of a few pebbles pushed together into a heap. What is important to the birds, in this case, is not the *nest* but the pebbles. They steal from each other in an endless round of larceny and the males present the pebbles to prospective mates with as much ceremony and intensity as if they were diamonds. As a matter of fact we have a lot in common with the Adelie penguin when it comes to courtship behavior. There are times when they seem almost as silly as people. Of course, that is only in appearance.

Pack Rat

"The wood rats get their other names (pack or trade rats) from their perfectly amazing passion for useless debris. They cannot resist shiny, oddly shaped, or otherwise mysteriously enticing objects. Just what a wood rat needs a button, empty cartridge case, or jacknife for may puzzle some. They use these treasures, of course, as building materials for their nests. . . . the wood rat is seldom found without a treasure tightly clamped in his teeth. He is almost always heading home with some woodland goody, and if he spots something even more enticing he will drop what he has and take the more attractive item. More than one camper has traded his wristwatch for an empty cartridge case without having a say in the deal."

>From
North American Mammals
by Roger A. Caras
Meredith, N.Y. 1967

Beaver Dam

The beaver builds the largest animal-made object in the world constructed by a small number of animals. A really huge termite nest may rival a beaver dam in weight but hundreds of thousands of termites are involved. Single beaver families (or a small group of families) have built beaver dams as high as eighteen feet. One that tall was found in Wyoming. Strangely, it was only thirty feet long. The Jefferson River in Montana once had a beaver dam 2,140 feet long. The all-time record appears to come from New Hampshire, however. One beaver dam found there was 4,000 feet long. The pond that backed up behind it held forty separate lodges. It was a veritable beaver housing development.

When we think of animals burrowing into the ground we think of worms and weasels, moles and other lesser creatures. But, other forms, too, seek the safety of earth surrounding them and cradling their young. Swift and certain hunters such as the wolf burrow deep to house their infant young and even a sky darter like the bank swallow, as the name implies, digs deep into a vertical face.

There are, of course, many benefits enjoyed by the burrowers. Temperature and humidity are more likely to be constant in the earth than in the air with its fickle give-and-take. It is quiet in the earth, and with only a single avenue to regard or defend, it is a little more restful for an alert mother. These are reasons enough but there are many more. In total, the earth is a nice place to be when you want to be alone, safe and quiet.

Wolf and Pups at Den

**Steep Clay Bank
Showing Burrow Entrances
of Bank Swallows**

Robin

Even those creatures we take for granted, birds like the robin, for instance, may have a quiet streak of architectural genius built into them as surely as the color of their plumage or the lilt of their territorial song. It can be very instructive, sometimes, to stop and look at what the common creatures do. They are so easy to miss. And we are so good at missing things.

"The nest (of the robin) is saddled on a horizontal branch or built in the crotch of trees of almost any kind, and it is commonly placed on the top rail of a fence, often on stumps, and, in fact, in all sorts of curious places, even in bird boxes. Orchards and the shade trees along streets are favorite nesting sites. The nest is a large, coarse structure, made of twigs, roots, stems, grasses, dry leaves, hair and wool. It is strengthened by a neatly made cup of clay or mud, which is surrounded by these materials."

      From *Nests and Eggs of North American Birds*
      by Oliver Davie
      The Landon Press, Columbus 1898

Wasp Nest

There are beauty and tragedy in the complex society of the wasp. That handsome structure, the wasp's nest, has a life of only a single season. By the end of the summer the wasp colony, like an ancient civilization, has begun to decay.
The small animals, confused by the impending tragedy, move around in a daze and make half-hearted attempts to repair damage to their nest or go through the motions of adding new partitions. It is all futile. Soon they fly away and never return or slip down into some secret corner of the nest and die. Only the mated queens will survive to hide away the winter under a rock or behind some bark. The following spring each will start the cycle over again. There are times when it would be nice to be able to take Mother Nature by the shoulders, shake her and say, "Now, see here. . . ."

Tsavo East Africa Termite Nest on Tree

"If you destroy a termitary you find firstly a tough resistant skin all around it. Under this skin you find that the whole termitary consists of cells through which a living stream constantly circulates. As you go deeper you find large passages and eventually a hollow, partly or entirely filled with more cells . . . if you go deep enough and observe carefully you will find at the very bottom a passage which goes right into the earth . . . it is the canal by which the termites get their water supply. . . ."

From *The Soul of the White Ant*
by Eugene N. Marais
Methuen & Co., London 1937

Trap-Door Spider Nest
The trap-door spiders dig tunnels in the ground and most species close them with a *hinged lid*. After the tunnel is completed it is lined with earth and saliva. Inside that a silk layer is added. The walls are then so firm that they retain their shape even when the surrounding ground is taken away. But, the remarkable thing is that lid. Here is an animal structure with a moving part. What should we call it?
Is it a place? A thing? Or a machine?
I think it is a little of all three.

Dew on Spider Web

"Two quite different mental pictures are normally associated with the words 'a spider's web.' One reader will see a rugged mass of dirt-laden silk in the corner of a room, and will experience a justifiable dislike of the animal which thus adds to her labours as housewife; while another will see an orb-web hanging in glittering perfection in his garden and will remember his wonder that so beautiful an object should be the work of so small a creature."

    From *The Spider's Web*
    by Theodore H. Savory
    Frederick Warne & Co., Ltd.
    London and New York 1952

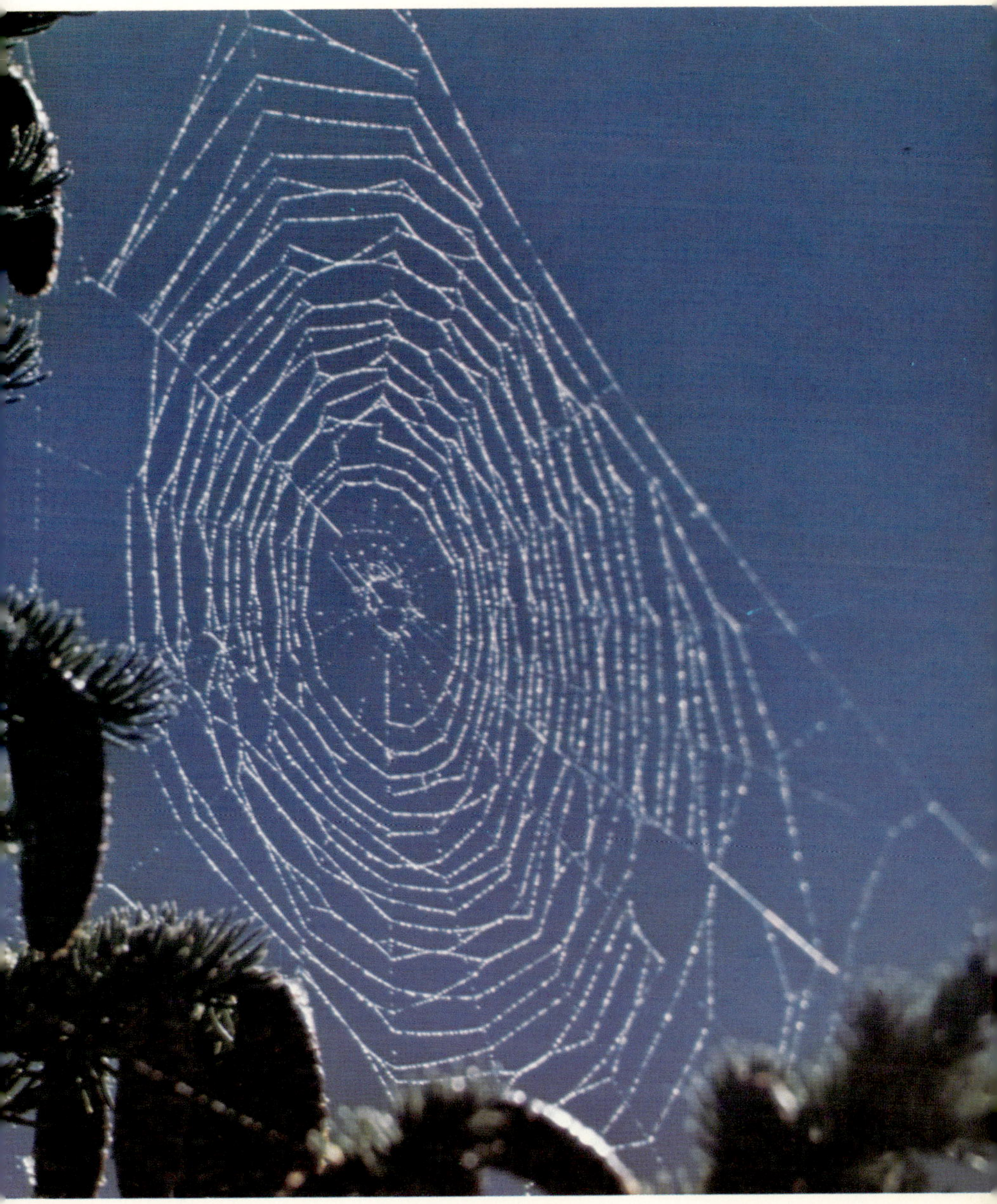

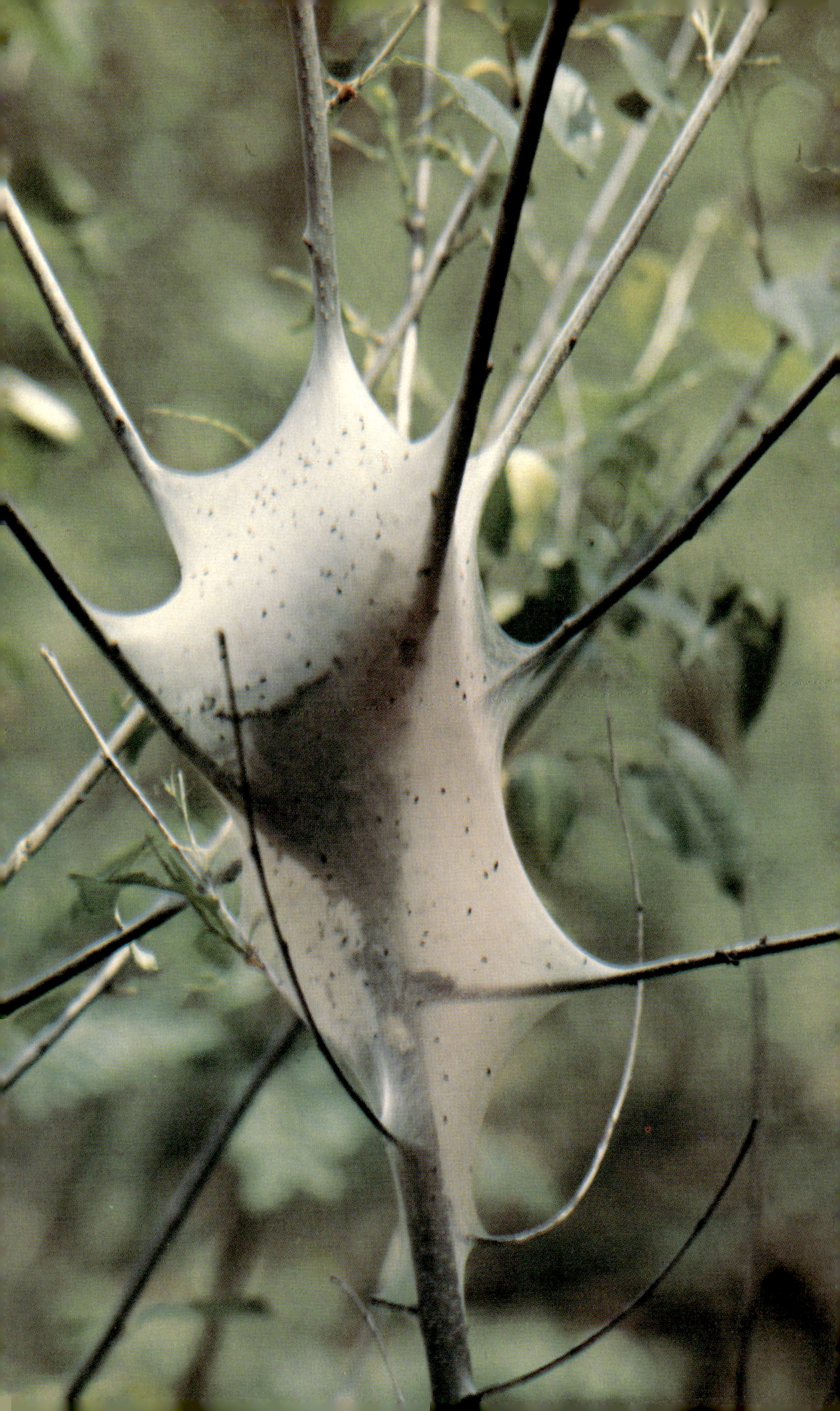

Tent Caterpillars
Caterpillars, moths and butterflies are not generally communal animals. But, the destructive caterpillars of the genus *Malacosoma,* at least some species, do spin a rough and random silk tent to hide in during the day. Masses of the little caterpillars gather inside as each adds his strands of silk. They emerge at night and in time, after repeated strippings, can kill the tree that is their host. It isn't really a society we are witnessing here, just a temporary, communal, silken quonset hut. We would probably consider them interesting animals if they weren't so very destructive.

Silk Moth Caterpillar
"Despite the invention of more or less silk-like plastics, silk is still used in large quantities. To produce twelve pounds of raw silk, growers feed thirty thousand worms one ton of mulberry leaves. . . ."
>From *The Insects*
>by Url Lanham
>Columbia University Press
>N.Y. 1964

Mourning Cloak Chrysalid
"The chrysalids are often irregularly shaped and studded with spines and tubercles. They frequently are very brightly colored, sometimes with brilliant metallic spots or tubercles; and from this fact came the name 'chrysalis' (chrysos being the Greek word for "golden") that is often used for butterfly pupae."
>From *Living Insects of the World*
>by Alexander B. Klots and Elsie B. Klots
>Doubleday & Company
>Garden City

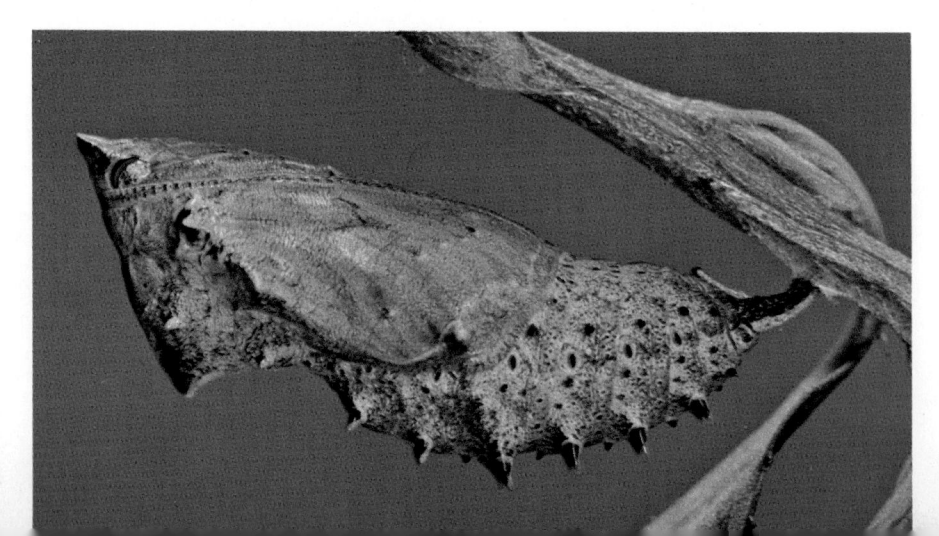

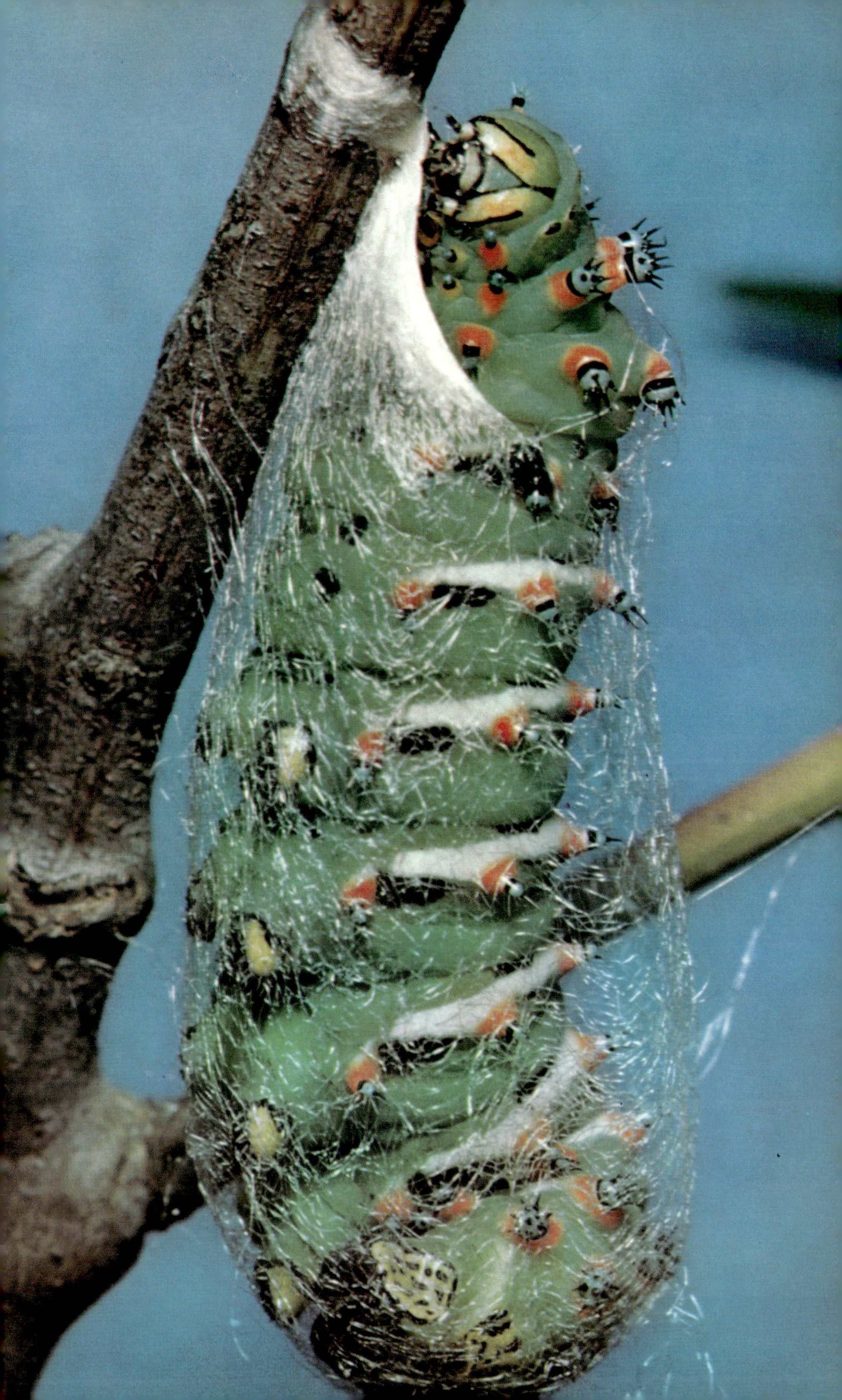

42

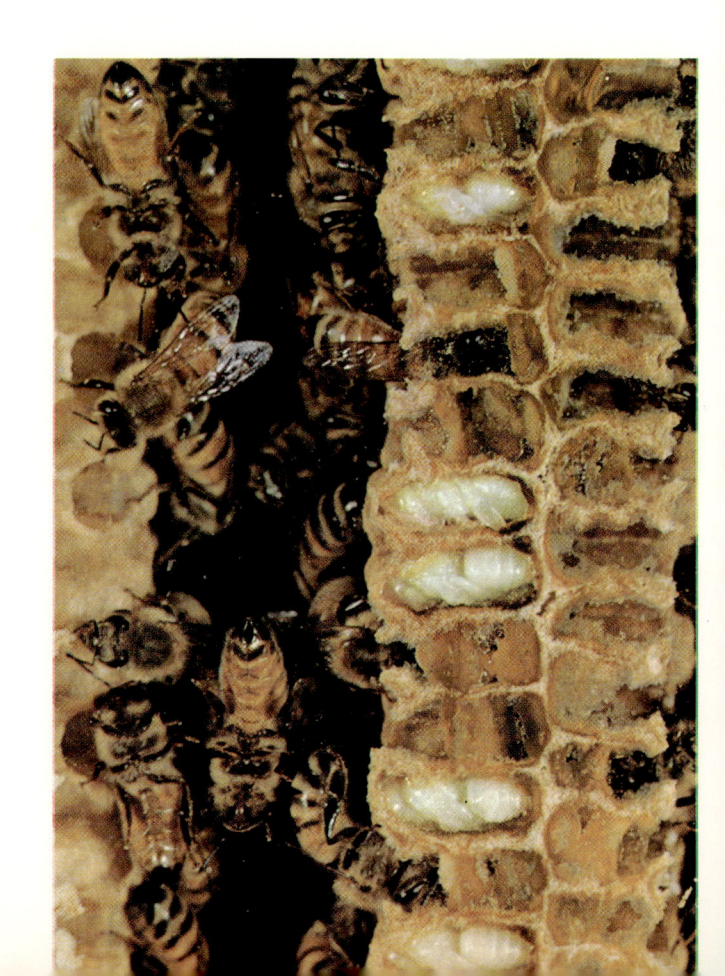

White-Browed Sparrow Weaver

The white-browed sparrow-weaver is a conspicuous species of southern Africa's dry, western Acacia area. In a large tree a single pair of these hyperactive little architects may construct as many as a dozen nests. Each bird sleeps in a different nest. Those nests made for roosting, as opposed to those used for the incubation of eggs and brooding of young, have two entrances. The birds sleep on a perch across the center—safe from interference and predation.

Sealed Larvae

Of all the truly social insects none is more meticulous in adhering to traditional design than the bees and wasps. The larvae are given their own sealed chambers and they are anxiously watched over by the workers whose life work it is to care for the young. There are no emotions involved, as far as we know, but the intensity with which these structures are created and guarded makes one wonder. There must be something in the instinctive behavior of these animals that does the same work as our emotions. That much devotion must have a qualitative as well as a quantitative drive behind it. It is more than just a *strong* instinct.

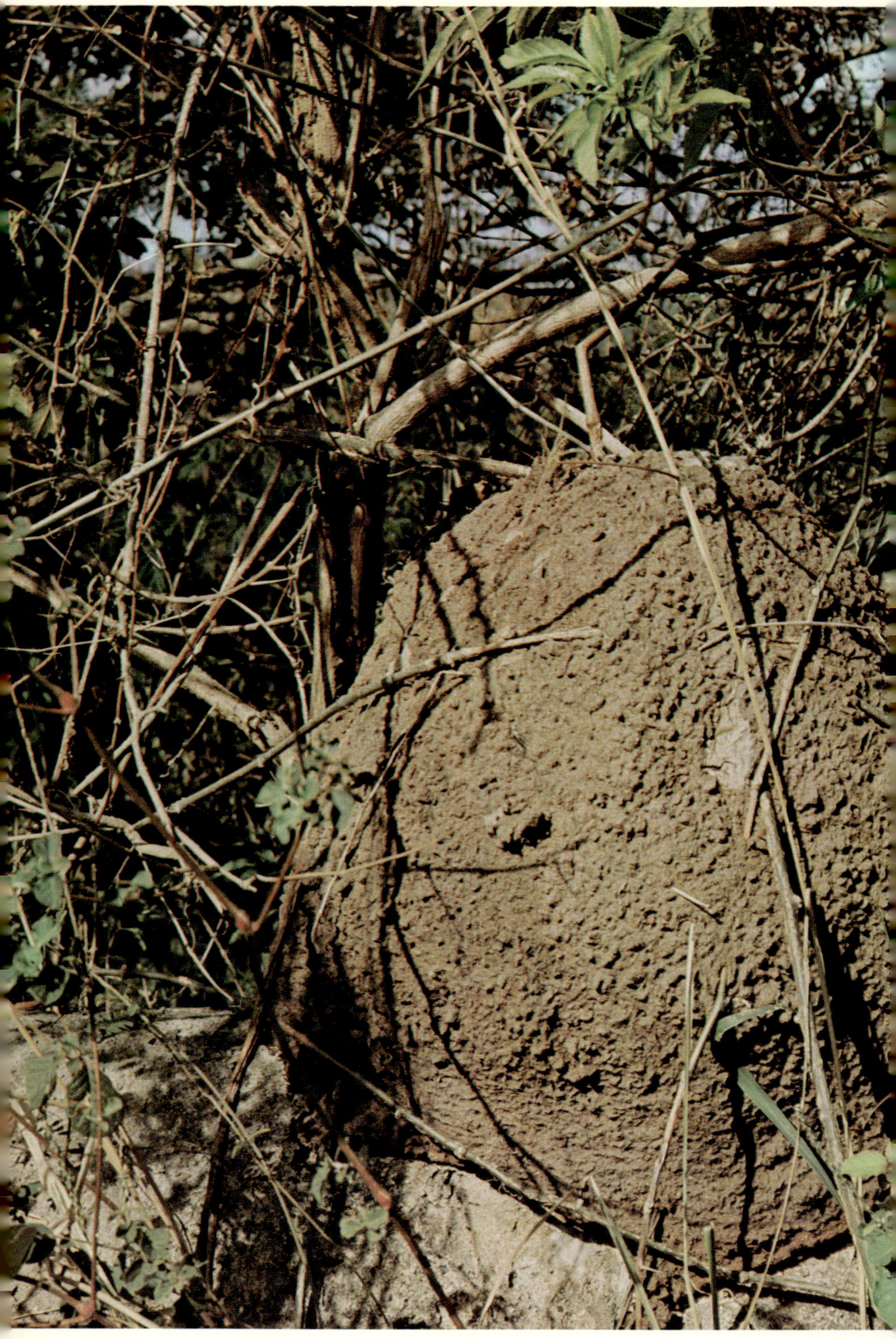

Ants

"In its essential features the typical (ant) nest is merely a system of intercommunicating cavities with one or more openings to the outside world. . . . The . . . cavities may be excavated in the soil or in plants. . . . The irregular form of the cavities is a characteristic so universal in ant nests that it would seem to be preferred to a monotonous regularity. It may be important, in fact, in enabling the ants to orient themselves readily."

> From *ANTS*
> by William Morton Wheeler
> Columbia University Press
> N.Y. 1910, 1965

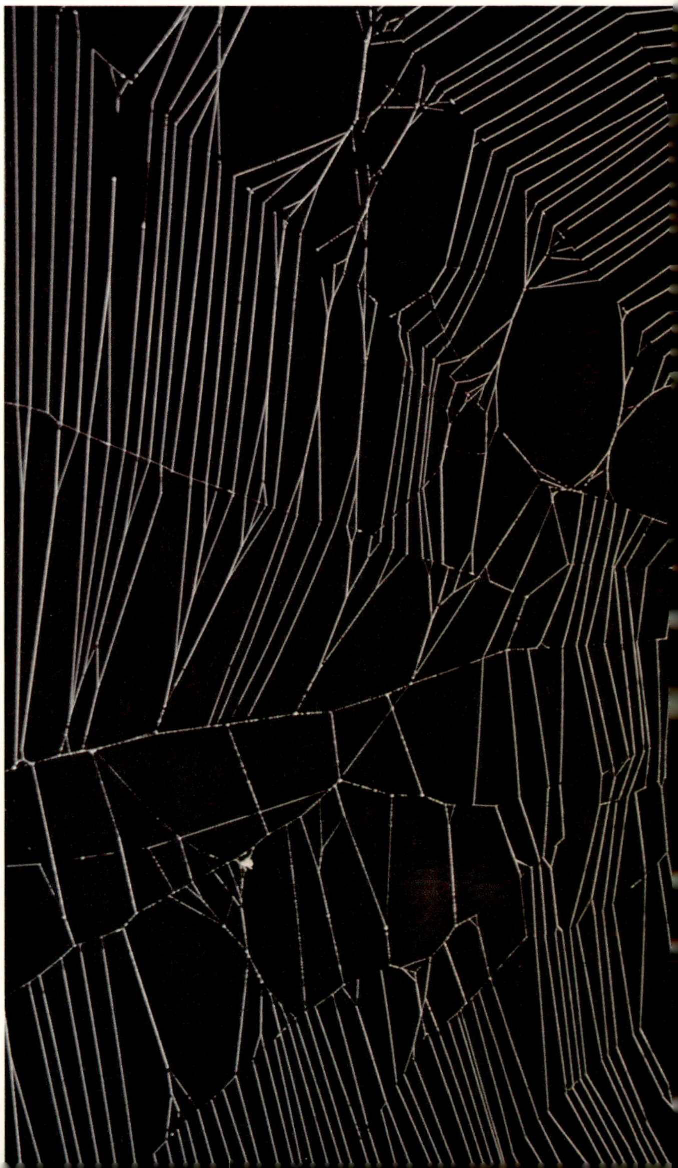

Orb Spider

It is difficult to imagine anything more perfectly wonderful than an orb web laden with dew, or catching the slanting rays of the sun. To say it is a string of jewels is to say the obvious, to attempt to really describe it is a fool's errand. As remarkable as its beauty is its strength. We have no steel as strong as this silk. And no communications network more perfectly designed to announce the intruder. An orb web is a structural miracle, an aesthetic high point on Planet Earth and a pre-electronics communication system of exquisite sensitivity. And the spider that builds it can appreciate none of these things. It is just programmed to build its web the way you and I are to scratch when we itch.

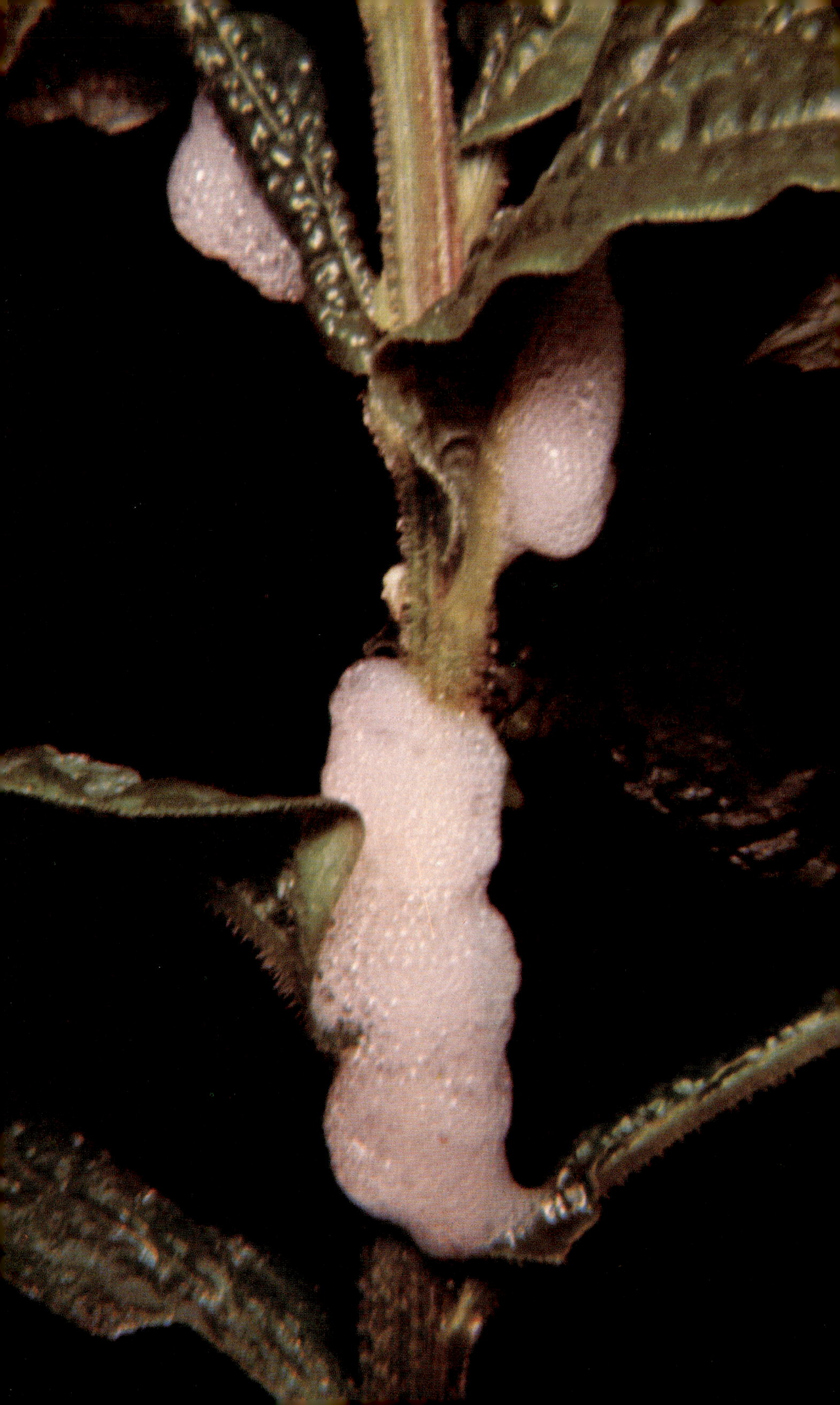

Spittle bug

"The young of the spittle bug live hidden in frothy masses produced by mixing air with a secretion that is poured from the digestive tract. Presumably this material affords protection from predators and parasites. So completely are the young spittle bugs adapted to life in this shelter that they die if removed from it, their thin skin being unable to prevent water loss."

       From *The Insects*
       by Url Lanham
       Columbia University Press
       N.Y. 1964

Nesting Osprey

Spotted Shag on Nest

A good many birds nest in colonies, or at least in traditional breeding areas. The ospreys build their nests in traditional areas near good fishing grounds. They seem to prefer exposed platforms atop poles and pillars, on the girders of radio towers, and on top of other clean, unbroken uprights. They will use a tree if they have to. An osprey nest can be huge after years of use. It can be a four-foot-thick platform of sticks and twigs and heaven alone knows what else.

The spotted shag also will use a tree if it has to although it prefers rocky ledges. Its nest is smaller and usually consists of seaweed along the coast and lesser vegetation on inland bodies of water. The shag colony is very much a reality and populations can be dense at nesting sites.

Water-Spider

Water spiders are altogether remarkable creatures. They construct a silken platform under water, attaching it to aquatic plants with silken strands stronger than tool steel. They carry bubbles of air down and release them under the canopy that soon billows out into a regular "diving bell." They even lay their eggs and see their young hatch out in their underwater hideaways.

Coast Horned Lizard

For animals like the horned toad (really a lizard) sand is many things. They not only burrow below for long-term protection, they use it as a short-term cover. Some desert snakes do as well. They "squiggle" themselves down into the sand in a matter of seconds and it is like pulling a canopy across. They become quite invisible in no time at all. In this kind of burrowing, feet are not used as diggers. The whole animal sort of undulates itself into the sand.

The "Stork-Town" of Austria

"No one can have travelled through Germany, in the summer season, without having had their interest excited by observing this the favourite bird of the country. In Holland, Belgium, and the German States, the stork not only meets with the kind protection afforded to all the feathered race, but is courted with a cherished and superstitious welcome, which seems to elevate it to a place amongst the household gods of the people. Invitations are held out to it by the inmates of the different villages to make their house its home; baskets of wirework, and boxes of wood, are erected on the roofs of some of the houses for the reception of its nest; and happy, and in good luck, is that person accounted, whose roof-tree becomes the object of their choice. It is a beautiful sight to watch these graceful birds, when they have young ones, standing, as is frequently their wont, statue-like amongst them, and as immovable as if they formed a part of the building."

From *Coloured Illustrations of the Eggs of British Birds*
by William C. Hewitson
John Van Voorst, London 1856

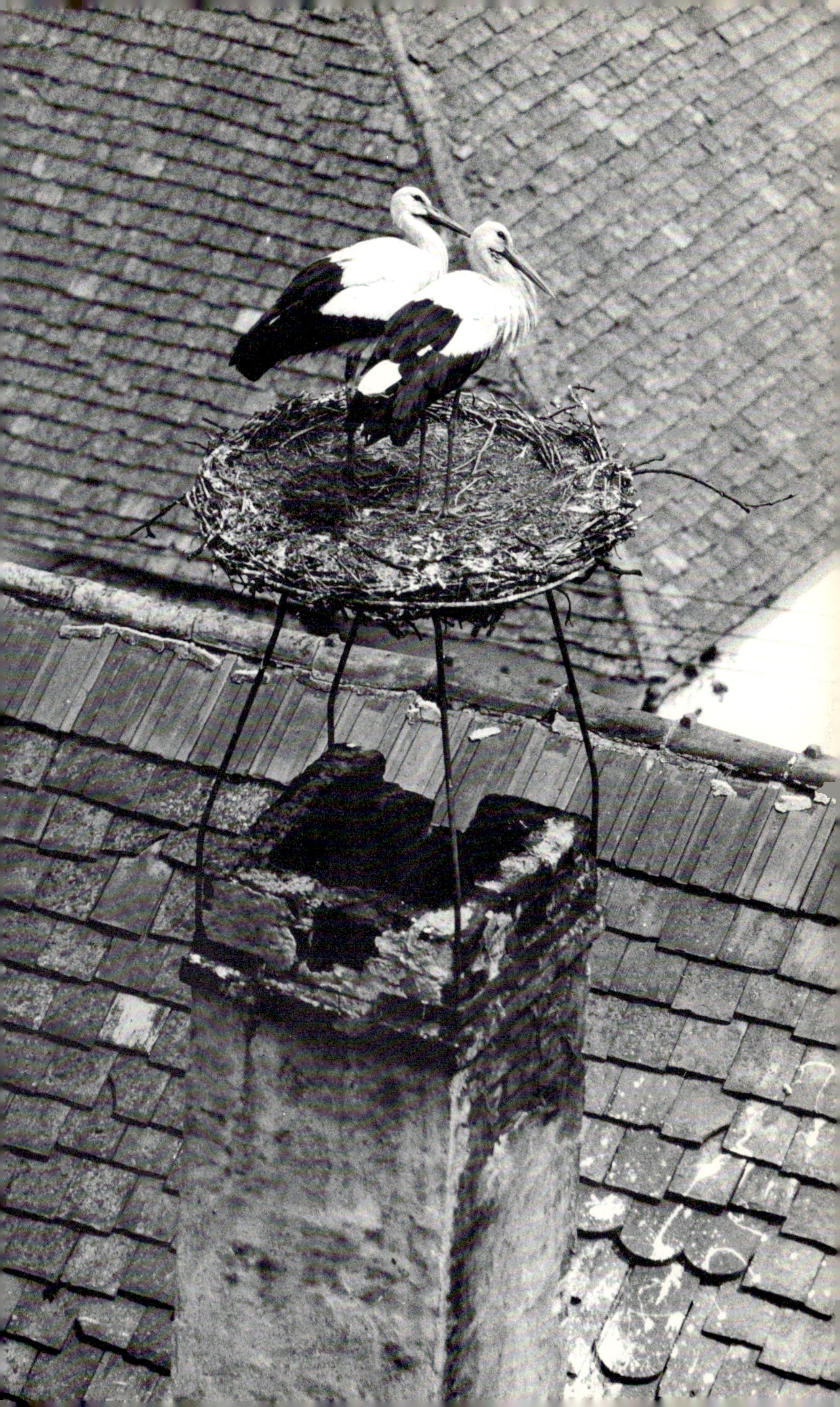

Pipe Organ "Mud Dauber" Wasp

A number of different wasps have developed special skills for working with mud. A variety of structures have evolved over the millennia like the work of the Mud Dauber, right. Wasps get their mud in two different ways. Some collect it from the banks of ponds or streams, and from puddles, while others carry water and mix it with earth at or near the building site. The two skills are very different but both have evolved to accommodate the life style of the little architects. It is interesting to contemplate, since the wasps that use these two quite different techniques are solving exactly the same problem. They have a mixture of water and earth to mold into nests for their eggs to develop in.

57

Malayan Sand Crab & Castle
Many species of crab burrow into the beach above or just below the high water mark. They are very highly adapted to a semiterrestrial way of life. They work endlessly at repairing and tidying up their small sandy domains. Each burrow is an established territory. They chase each retreating wave down the slope of the beach, looking for morsels of food and dash ghostlike back to their burrows on the approach of new water. Those that construct their burrows below the high water mark plunge below just in time to avoid being swamped. On islands in the Indian Ocean I have watched for hours as the beach exploded into activity as each rush of water reversed itself and headed back for the sea. Then, as each new wave broke, the activity would double in intensity for a second or two, and then all would be still until the water started back again. It is an endlessly fascinating panorama. It made me wonder what we would look like to other beings as we plunged into our subway entrances at the beginning of the rush hour.

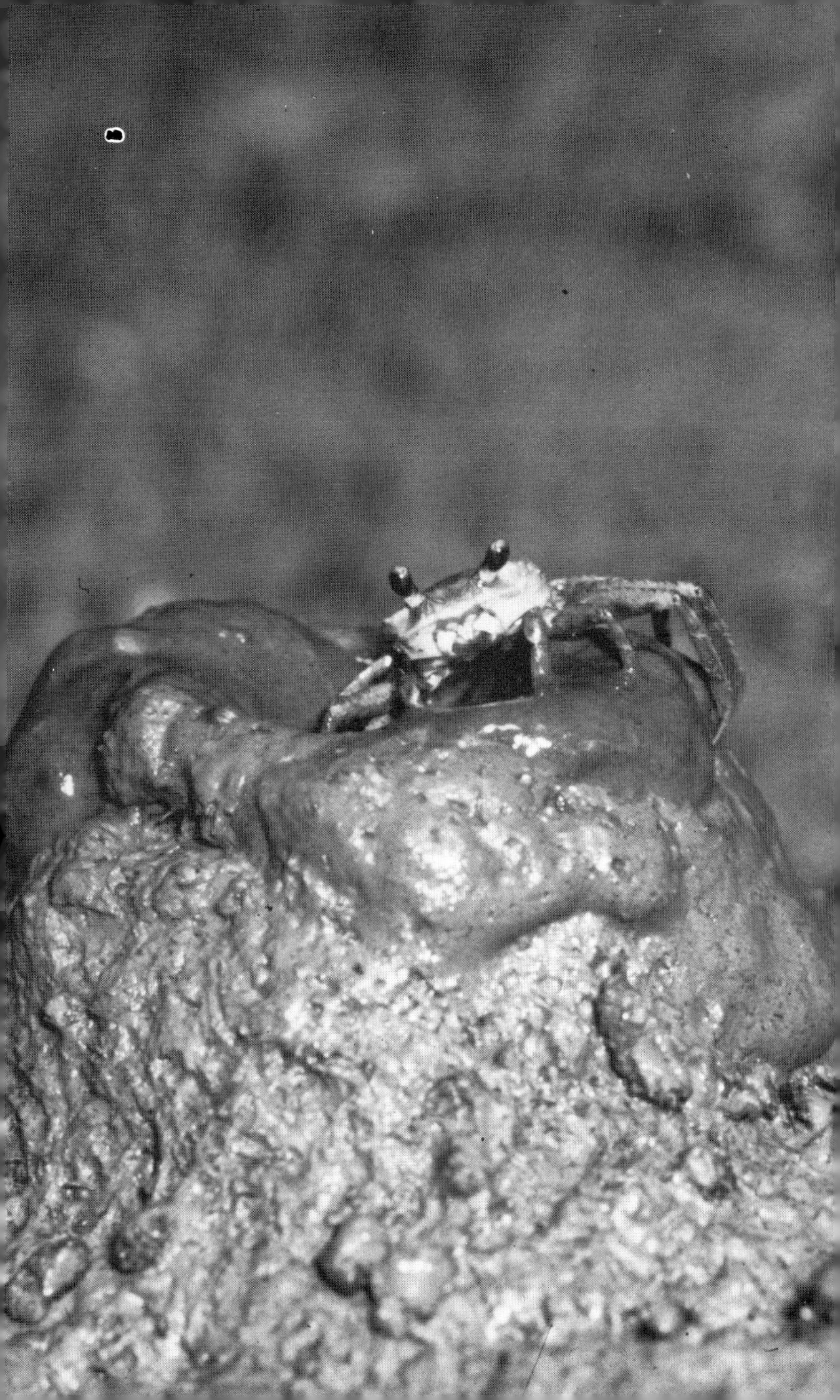

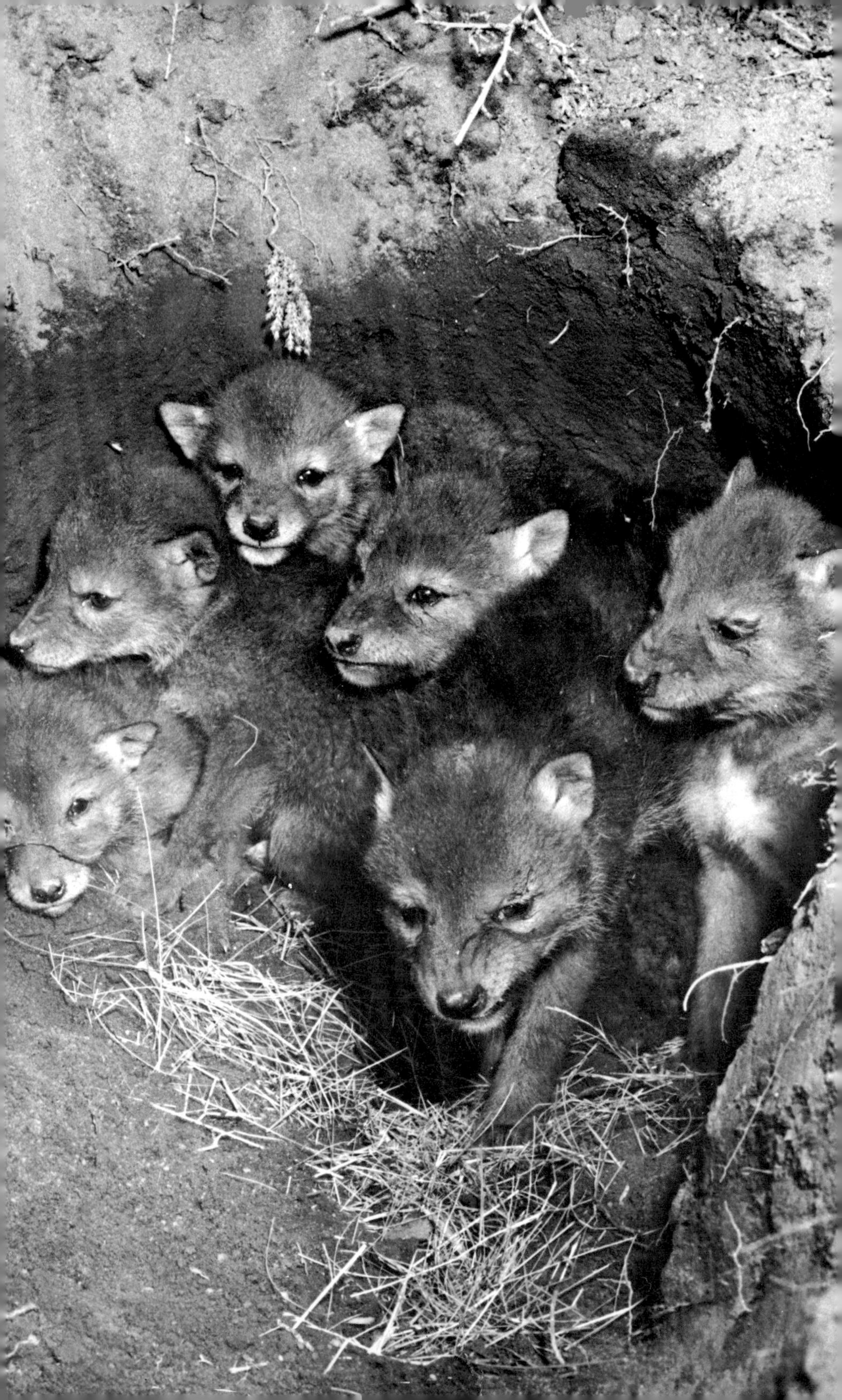

Coyote Litter

"The female (coyote) continues digging and cleaning out den holes, sometimes a dozen or more, until the young are born. Then, if one den is disturbed the family moves to another. Sometimes the animals move only a few hundred yards, apparently just to have a cleaner home, leaving many fleas behind."

      From *The Clever Coyote*
      by Stanley P. Young
      Part I
      The Stackpole Company
      Harrisburg, 1951

Raccoon

The raccoon is the cartoon character of the animal kingdom. Yet, raccoons are very real creatures as well. When you see them peering out of their nest in the hollow of a tree they seem to be defying us. "That's as far as you go," they seem to be saying, "this is where our secret world begins." Indeed, that is what animal architecture is all about—it is the way in which animals keep their secrets. And in the animal kingdom a secret is a means of survival.

PHOTO CREDITS

PAGE	SUBJECT	PHOTOGRAPHER
5	White-Footed Mouse	Karl H. Maslowski
6	Bird at Nest	Toni Angermayer
8	Elf Owl in Cactus	Lewis Wayne Walker
9	Wolf Spider	Ron Winch
10	House Martin at Nest	Russ Kinne
10	Weasel	Karl H. Maslowski
11	Stripped Ground Squirrel	Des Bartlett
12	Little Brown Bats on attic wall	Karl H. Maslowski
13	Red-Eyed Vireo at nest	Karl H. Maslowski
14	Chipmunk	Russ Kinne
16	Soft Corals	Russ Kinne
18	Fairy Tern	Lewis Wayne Walker
19	Flamingo Nest and Egg	Porterfield-Chickering
19	Black Swans and Nest	Eric Lindgren
20	Badger	J. Behnke
21	Galapagos Tortoise with Nest and Eggs at Burrow	Miguel Castro
22	Adelie Penguins	George Holton
24	Pack Rat	Russ Kinne
26	Beaver Dam	Russ Kinne
28	Wolf and Pups at Den	Russ Kinne
28	Bank Swallow Nests	George Whiteley
30	Robin at Nest	Karl H. Maslowski
32	Spider Web	Olaf Soot
34	Boabab and Termite Nest	Philippa Scott
35	Trap-Door Spider Nest	Louis Quitt
36	Wasp Nest	Syd Greenberg
38	Tent Caterpillar	R. V. Boger
40	Mourning Cloak Chrysalid	Louis Quitt
41	Silk Moth Caterpillar	James Carmichael
42	White-Browed Sparrow Weaver	George Dineen
42	Sealed Larva Bees	Treat Davidson
44	Ant Nest	Fritz Henle
46	Orb Spider and Web	Syd Greenberg
49	Spittlebug Nest	Jane Kinne
50	Nesting Osprey (8H3093)	Richard Frear
50	Spotted Shag On Nest	Chuck Meyer
52	Water Spider	Lawrence Perkins
53	Horned Lizard	Verna Johnston
55	Storks on Chimney	Andy Bernhaut
57	Pipe **Organ** Mud Dauber Wasp	Karl H. Maslowski
59	Malayan Sand Crab	Jane Burton
60	Coyote Pups at Den	Joe Van Wormer
63	Raccoons at Den	Karl H. Maslowski